The cookbook of comfort

Jim Patterson

29 Rattle Hill Rd.

Southampton, MA 01073

www.orchardvalleyhc.com

Email: jimp@orchardvalleyhc.com

authorHOUSE®

AuthorHouse™
1663 Liberty Drive
Bloomington, IN 47403
www.authorhouse.com
Phone: 1-800-839-8640

First published by AuthorHouse 10/08/2009

ISBN: 978-1-4389-5990-0 (sc)

Library of Congress Control Number: 2009910795

Printed in the United States of America
Bloomington, Indiana

This book is printed on acid-free paper.

Introduction:

This book spans more than the basic "how to guide" of residential comfort systems. A home's comfort system can take so many potential paths. Choosing the right one for the application is the key to success. My vision is to create awareness beyond the mechanical requirements of the installation process and develop an understanding of *why* we do what we do. There are dozens of books reviewing the "connect pipe a to pipe b" approach without ever starting at square one, why we need the pipes and what they do. You will notice that I immediately transform your title from reader to "aspiring comfort expert" as you begin these pages of text. If you are taking the time to read this book, your interest goes beyond that of a casual reader and your normal title of homeowner, architect, builder or hvac tech will be removed and upgraded. Even if you are a homeowner looking for more insight on your new home's comfort system options, the knowledge this manuscript provides will elevate your knowledge above that of most contractors you will interview.

The pages you are about to enter are based on the past two decades of my experience in residential home comfort system design and installation. My reputation in my territory is the "go to" guy when there is a complex design, mechanical problem, or high quality installation. Most of my projects are attracted by word of mouth and that, in my mind, is the greatest form a praise a tradesman can achieve. Sure, this may be a volume of my personal opinions, but so far, they have created an amazing business that grows by leaps and bounds every year. Selling comfort is an art and without proper artisan training, you always are one step short of excellence. I am hoping my "cookbook" will expedite your journey and provide a shortened path to enlightenment!

All the best,

Jim Patterson

Table of Contents

"The sweetness of low price is
soon overshadowed by the
bitterness of poor quality"

Chapter

1

Defining the comfort expert:: a view from the trenches

What a unique opportunity the business of residential heating and cooling presents to mechanically inclined individuals. The homes in which we work provide a blank canvas, awaiting an artisan's hand to weave the ducting and pex tubing through its structural members. Every client presents the opportunity to showcase our talents, by assembling a home comfort system that is custom tailored to their lifestyle and home design. Leaving our unique trademark on a client's home is like an artist signing their masterpiece, proudly announcing, "I did this!"

The quest for comfort is a difficult trail to maneuver. Whether you start this journey as a homeowner, builder, architect or HVAC professional, the path is often poorly marked and getting lost is pretty common. Understanding the conditioning needs of each project is the first step to proper design. What works in one home is not necessarily the best choice for another project. Knowing what is required for comfort and how each home's environment differs is one of most critical skills you could perfect. Understanding these demands will allow an intelligent and informed conversation on each project's HVAC requirements and help avoid the pitfalls that plague most installations.

The next phase will be the realization of cost and allotting a realistic budget to your application. This is one of the hardest parts of any project as the price ranges will certainly add confusion and sometimes aggravation and disappointment. Unfortunately, the construction climate is primarily driven by price, the lower the better. It is often difficult to explain why one price is more than another and how the designs and equipment differ. After all, HVAC is HVAC, isn't it? Not exactly, all systems are not created equal and having the knowledge to strain the grounds from the coffee will help you make a more informed decision when the time comes to choose your "artisan". I see far too many projects take place based on a price alone and when asked what that price delivers, I usually get a blank stare. Perhaps this is where the realization occurs that more reasearch should have taken place. Many times a decision is rushed by a demanding schedule and time is not taken for the proper ground work before the trigger is pulled. Regardless of how the wrong decision was made, the fact it was made at all is a crime. As a heating professional, my job is to prove to the consumer that my design will satisfy the job's requirements and why my assembly of equipment is their best choice. The burden of education lies on me. I never want to be the low bidder, not because I want a Porsche (I actually had a homeowner ask me what kind of Porsche I drove after handing over the quote. Despite the initial sticker shock, I was hired to install their new home's comfort system after the realization that my system was the best value) but because quality costs more and I will provide a better system and back it up properly. You must decide how much of a chance the low price is worth to you and if you can afford to replace / rebuild it if it falls short of expectations.

Expectations- the second driving force behind your decision. What type of performance are you willing to settle for. If you simply want hot air and cold air, you can get away with less and pay less. Expecting comfort, efficiency and longevity adds to the equation and the costs rise in relation to the degree of perfection you prefer. Everyone wants to pay less for their home conditioning. To achieve a lower ulitity bill you need to insulate better and buy a more efficient system. These improvements add cost now but pay for themselves over years of operation, one of the few options that offer that luxury. Your lifestyle and home design also demand certain details from your comfort system. Tight buildings must have proper ventilation, multi story homes need zoning, occupants with allergies need excellent air filtration and tile floors need radiant (not exactly, but they sure are more comfortable). Understanding the options and knowing when they are needed will help choose the correct design for your application.

The final difficult decision is a choice in contractors. You are entering an agreement with a company or individual that should last as long as you own the home and beyond. A quality contractor is hard to find and once you know you have made the right choice, you should be able to enjoy this facet of the project knowing all of the correct decisions have been made and eagerly anticipate the unique environment about to be created.

NOTES

The cookbook approach to comfort

Designing successful HVAC systems is much easier than it may initially appear. As long as we keep a few simple rules in check and follow the design up with a quality installation, comfort is assured. The design approach is best perfected when compared to the preparation of a meal. The problem with many HVAC systems is that the recipe is incomplete, a crucial ingredient left carelessly aside. Follow the recipe, and the meal is a success. My cookbook approach to making a client's home comfortable has been the key to success. The guidelines make it easy to approach every home as an empty plate and choose the meal based on the structure's unique demands. Not everyone's tastes are the same and expanding your cooking repetoire is essential. The following pages are the foundation on which every system is built. Start with these concepts in mind and adapt them to your personal and home environment demands. With the proper blend of ingredients, a unique flavor of comfort will be created and savored for years to come.

Learning your way around the kitchen.

There is little difference between preparing a full course meal and designing a home comfort system. The "chef" combines the ingredients and the result should be worthy of a five star rating. The key is to understand which ingredients will determine successful indoor conditions

Before we get into the recipe, we need to analyze the ingredients, or the *four factors of comfort.* These four items are the heart and soul of every design and identifying how to adapt them to the home's structure is vital.

1. Temperature control

2. Humidification

3. Dehumidification

4. Air quality

There are numerous sub categories under each factor, however these four are the main building blocks we analyze to start our design. Another important task is to learn any regional comfort issues you may have in your corner of the world. Here in New

England, I see every side of good old Mother Nature, from freezing winter days to oppressive summer heat and humidity. I must look at every factor and ensure it has been addressed in the design concept. Knowing the climate is as important as knowing how to assemble the final system. What works here in New England may not suffice in the Southeast or the Mid-West and identifying the sometimes subtle demands is imperative.

My recipe for a home comfort system

I saw a sign in one of my client's offices that read, *"The sweetness of low price is soon overshadowed by the bitterness of poor quality."* Although budget requirements drive everything we do, try to avoid a hasty decision based solely on capital expense. These systems can be built in phases, adding components as budget allows. The following list will give you a few items to ponder while contemplating your HVAC system requirements.

- What kind of fuel will be used to heat and cool the home?

- Is central cooling a major priority?

- What kind of temperature control would you like? Many homes have specific demands created by design and orientation. Zoning difficult to condition rooms is one aspect of proper design that is often left off of the shopping list.

- Are there any large use items for domestic hot water? Large tubs, many people...

- What do you want to see (or not see) for heat emitters in the rooms? Grills, baseboard...

- Humidity is a critical aspect of winter home conditioning. Pay careful attention to this detail.

- Are there any allergies or asthma suffers in the home?

- Hot water boiler vs. warm air furnace vs. geothermal vs. hybrid? This option sets the stage for the rest of the options. See the following pages for details.

- Has a budget been established for the home's comfort system? (This is imperative to ask as it seems everyone has a misconception that a home's heating and cooling system is 2% to 4% of the home's total cost. I find that 8% is usually the minimum, and cost escalates with features)

- Is future expansion of the living space a possibility? Simple design changes now may reduce costs later.

- INSULATION SYSTEMS- research the options you have to insulate the walls ceilings and floors of your home. Spend as much here as you can afford as a better building envelope will reduce the size and cost of the HVAC system and minimize your annual conditioning bills.

- Radiant heating- a nice luxury would be warm floors. Take the time to evaluate this alternative

- Bath ventilation- many new systems combine bath exhaust with fresh air exchange and well insulated homes need a source of mechanical ventilation.

- Sizing- what capacity should the system be? A careful heat loss/gain calculation is the basis for EVERY design. Rule of thumb anecdotes have no place here.

A good example of bad ducting. Avoid this disaster at all costs

"filter bypass" ducting technique. The air is meant to go through the filter, not around it!

My recipe for success

The four factors of comfort
1. temperature
2. humidification
3. dehumidification
4. air quality

Address all of these with your home's
HVAC system to assure a comfortable
environment in every season

Ingredients of a comfort system

ATTENTION TO BUDGET CONCERNS

The first task of any design is to carefully analyze the future occupant's expectations of the system's operation. What functions are the most important? What bonus features would be nice additions? Several basic comfort questions should be reviewed to establish a direction for both present and future demands. Short-term sacrifices to meet tight budgets can lead to expensive modifications down the road or unsatisfactory comfort for years to come. Try to plan for all comfort needs now, and at the least, get the basics done and defer the installation of the optional major components until later. Your heating and cooling bill is a recurring expense, and installing a more efficient system will pay for itself over time.

FURNACE VS. BOILER VS. GEOTHERMAL VS. HYBRID ...

This decision is based more on budget demands than feature and design benefits. A warm air furnace is more cost effective initially and can provide a comfortable environment if designed and installed properly. Duct systems allow us to deliver comfort year round, and add the benefits of air quality improving products. Humidifiers can be installed to introduce moisture during the winter months, ventilation systems can supply the duct system with fresh air to distribute throughout the home, ultraviolet light systems can rid the air of bacteria and molds, and HEPA filters can scrub the air to new levels of perfection. The effectiveness of all of these components lies in the air handler chosen to move the air through the ducts.   Variable speed, multi stage gas furnaces are the best options when a standard fossil fuel warm air design is called for. These appliances can be coupled with multi-stage exterior condensing units or heat pumps and zoning systems for a complete comfort package.

Boiler based systems offer a much broader range of "bonus" options when compared to a more standard furnace system. The boiler simply acts as the "BTU plant" making the heat to be distributed through the home in any manner the budget allows. The beauty of this system is the variety of options available to improve the comfort of the occupants.   The diversity of design adds value to the system, allowing the homeowner the opportunity to select the comfort options from your HVAC menu. Radiant floors, European wall panels, snow melting, pool and spa heating, baseboard, domestic hot water production, and hydronic warm air systems fill out the spectrum of heating options when a central boiler system is selected. Boiler based systems cost more than fossil fuel furnaces, but given the opportunity, they can provide unsurpassed comfort and reliability. One final note regarding the boiler or furnace you select. Remember this is the heart of the home's comfort system. The longevity and efficiency of operation all fall back to this single appliance. Invest wisely at this level; saving a few dollars here on a cheaper brand or less sophisticated model may cost valuable dollars over the long run.

As technology evolves, many other advanced systems are jumping into the game. Geothermal is one technology that is making a name for itself with the "green" focus.   Geothermal designs can be complex and require the attention of an experienced contractor. My personal preference is a closed loop system. This piping design features a series of vertical wells or horizontal trenches with a blend of antifreeze and water (sometimes alcohol and water) within the exterior loops. This fluid circulates with a pump system and uses an interior heat pump for energy transfer. The design can heat / cool the home year round and offer some assistance heating the home's domestic hot water. Water to water heat pumps can augment a radiant floor heating system, pool-heating, snowmelt... most anything we can do with a boiler in low and medium temperature applications. These systems have a hefty upfront cost and make sense on smaller super-insulated homes. Large homes or retrofits are tough to justify as the cost of the wells or trenching can be a deal breaker more often than not.

I am a big fan of the hybrid heating systems that are popping up during the recent fuel crisis. Hybrid designs feature heat pumps coupled with furnaces or hydro-air systems. They use the exterior heat pump as the first stage of heat and kick in the fossil fuel furnace or boiler when the demand is greater than the heat pump's capacity (as it gets colder outside, the heat pump's effectiveness drops off). The heat pump advances in recent years makes this alternative a fantastic option where geothermal is financially out of range. Many of the new systems are multi stage (small and large heat pump in one box) and use R410a refrigerant (green alternative to chlorine based R22).

There are even more base system options appearing every day and this facet of the design is one that should be a critical concern. Investigate all of these options to see which one makes the most sense in your home.

Now that the base system is set in stone and a design objective has been achieved, it is time to review the types of components that can be added to the system to allow it to reach your desired comfort plateau. The base system simply adds or removes heat from the home. The components added to the system provide additional comfort, protect your health and home, and even reduce the operational costs over the course of the year. The components you choose to add are determined by the following:

1. Base system (the existence of ducts within the home? Hydronic heat only or ducted design)

2. Health concerns

3. Home construction tightness

4. Budget limits

Comfort is a state of being that requires all four factors to be in sync and carefully tuned to the occupants individual preferences. The home's design also places demands on a HVAC system and addressing these issues is the key to completing the package. The end result will be a blend of equipment that will work in unison to make your home a comfortable, healthy place to live. The following pages review each comfort factor and how I use different strategies to control the comfort within a home.

Addressing the "four factors of comfort"

1. TEMPERATURE

The most important factor of home comfort is temperature. Our bodies prefer an environment that is thermally stable . Conditioning the home is a delicate subject, with a multitude of products and techniques available. The reviews below provide a quick overview of the possibilities.

Conventional duct systems (heating and cooling)

This method of conditioning the air is the most basic approach to heating, cooling, and ventilating your home. These systems feature furnaces or air handlers that are designed to heat or cool the air and blow it through the duct system. The ducting is generally made of sheet metal and can eat up a lot of space in the basement and attic. This approach offers the greatist variety of accessories due to the availability of the central duct system. The best system for air movement is an appliance with a variable speed fan. This component will provide a quieter installation (provided the duct system is designed properly) with more even temperatures, reduced summer humidity, and better effectiveness from the air quality components. These systems can feature zoning to allow more than one thermostat in the home, increasing the operational efficiency and comfort. Whether you choose a furnace or hydro air option, the variable speed fan combined with a zoned design is my preferred choice EVERY time. The variable speed systems are amazing and provide an unmatched level of comfort and accessories. (Yes, a perfect world would feature a bit of radiant floor within the home, as the comfort of warm floors under your feet is immeasurable). Planning room for the ducting is a necessity and difficult to address in some instances, but proper duct size leads to a quiet system, longevity and comfort. Address this concern at the start and less panic will ensue when a proper design will not fit!

Miniduct systems (heating and cooling)

These systems have a variety of uses, ranging from new construction to retrofit applications. The supply duct system is much smaller than the conventional counterpart is. This is due to the system's specialized "high velocity" air stream. The room grills that the air exits from are small 2-inch holes, greatly reducing the aesthetic impact. The zoning is typically achieved with individual systems, one air handler per thermostat. If multiple zones are required (multilevel homes), each floor must be served by its own air system. If heating is to be added (for heating and cooling applications), then a boiler must be used to create the heat that will be pumped to the air systems through small pipes. These systems are invaluable when older homes require cooling and space is limited for ducting.

When using a central boiler for heating, a variety of heated water accessories can be added to serve the home. The versatility of a boiler system certainly makes it the best long-term option (more expensive initially), and with the options below, offers the most comfort. These systems are popular in Europe and some portions of the United States (The Northeast is a Mecca for "wet heads") and is seeing a renewed growth in the States due to the popularity of radiant floor heating.

RADIANT FLOOR HEATING

This is an exciting system, featuring special plastic tubing that attaches above or below the flooring. These tubes have low temperature water flowing through them and keep the home flooded with a coziness that is not possible with any other form of heat. The radiant tubing can also be put in cement slabs (to condition a basement or garage), around tile shower walls, or even melt snow in sidewalks, stairs, and driveways.

PANEL RADIATORS

These are wall hung heating units that can provide wonderful warmth and add a number of temperature control options. They are typically made of stamped or welded steel and are available in a variety of shapes and colors. These panels can provide radiant comfort where floor heating is not possible.

BASEBOARD STYLES

These tend to be the most basic of hot water heating sources. There are several styles available, most feature a metal enclosure with an aluminum finned copper pipe core. They connect to the boiler with copper or pex tubes and usually extend along the outside perimeter of the room. These heat emitters rely on convection, which is a movement of cold air into the bottom of the radiation and heated air rising to the top. This creates a circular air current with warm air rising to the ceiling and cooler air drawn across the floor.

2. HUMIDITY CONTROL

The second factor of comfort is the control of humidity in a home, specifically humidity for use in the winter months. Humidity retention in the winter is the most difficult aspect of comfort to control. The home is usually more to blame than the HVAC system when the moisture levels drop. Coaching the owner on the importance of a tight home is of utmost importance.

How does humidity affect comfort? The human body is made mostly of water, an environment that is too dry will actually wick water from the skin and leave a chilled sensation. This leads to us turning up the thermostat and burning more fuel. Health concerns include dry and cracked skin, increased respiratory infections and worsened allergy symptoms.

The home also suffers from the moisture deficit. The wood products within the home shrink as the moisture is drawn from them, leaving gaps in floors and weakening furniture. Trim and moldings also show signs of shrinkage as once tight joints begin to gap and crack.

Now that we know we must have moisture in the home, what do we do next?

1. Build the house as tight as possible.

2. Install a variable speed air handler

3. Steam humidifiers are the best. They are more costly to install and operate, but serve their purpose perfectly.

4. Attic ducting in cold climates must be insulated well if humidification is to be used. Condensation can occur and lead to problems down the road.

1. Steam humidifiers: these are typically the best humidification systems available. They contain a steam canister inside the enclosure and inject a stream of steam into the duct for distribution through the home.

2. Power humidifiers: These units work on the principal of evaporation. The water trickles over a screen in the unit and a small fan blows dry house air over the pad, absorbing water as it goes by. These may be OK if the home is tight enough

3. Bypass humidifiers: work on the same principal as power humidifiers, using the air system fan to move the air.

3. DEHUMIDIFICATION

The third factor of comfort is the control of moisture in our homes during the cooling season. The purpose of our home conditioning systems during this period is to reduce the temperature levels and remove moisture from the air. To do this properly, the system must be sized accurately and the right appliances be put in place to mechanically cool the air.

A variety of options exist for the exterior cooling unit. Conventional condensing units are the norm, with efficiencies ranging from 10 seer up to 19 seer(the higher the seer rating, the less it will cost to use). A second option is a chiller. This device uses refrigerant to cool water through a heat exchanger within the cabinet. The fluid (water and antifreeze mix) is pumped to the air handlers for cooling capacity. The air handlers would use a chilled water coil then a conventional DX coil. These are good alternatives if a home requires multiple air handlers (especially miniduct systems)

Depending on the climate and home construction, some residences need stand alone dehumidification units connected to the duct system. These commercial grade appliances take air from the home, remove the moisture, and dump it back into the duct stream.

4. AIR QUALITY:

A SMORGASBORD OF COMFORT COMPONENTS

Last, but certainly not least, comes the topic of air cleanliness. A number of reasons exist to remove the debris from the environment we live in. The first is to create a healthy environment. The cleaner the air we breathe, the better it is for us. That is a straightforward assessment. Some individuals have reactions to airborne contaminants that pollute the air in our homes. Dust mites, pet dander, pollens, molds, and virus' all surround us as we live in our homes. Removing them can help mitigate allergies and asthma. A multitude of accessories can be added to your HVAC system to combat the contaminants that build up in your home. The overview below provides a view of the possibilities.

AIR FILTRATION

Removing particulate matter from the air we breathe is the most basic step to creating a pristine living environment. To improve the quality of the air, we must purify it by removing contaminants. Passing the air through a filter is the first step.

STANDARD PLEATED FILTERS

This filtration assembly is the first line of defense. Most warm air furnaces and air handlers include a standard one-inch thick filter. These stock parts collect the larger particles entering the webbing. They are great for stopping cotton balls, but little else! Many manufacturers offer replacement, high-density filters that require washing to remove trapped contaminants. These are better than the mesh style frames, but need service frequently to keep the airflow constant.

CABINET MEDIA FILTERS

These should be the standard filter choice on every system. The cabinet enclosure attaches to the furnace or air handler return and provides improved performance. These filters have large accordion style collector surfaces to extend the service life of the units by adding surface area. They require replacement 1 or 2 times a year in most homes.

HEPA FILTERS

These are the most advanced filtration devices used in duct system applications. These filters mount on or near the duct and draw a percentage of air from the duct and return the same amount after it has been cleaned. This configuration is referred to as a bypass installation. We typically use filters in series with a cabinet media filter. The media catches the larger particles and the HEPA snatches the smaller ones.

ULTRA VIOLET LIGHT AIR PURIFIERS

This is the final line of defense in your home's air cleaning system. The UV light installs in the system ducting and is used to "kill" living organisms in the air stream. We use this technology in unison with constant fan systems to scrub the air clean and provide the cleanest environment possible.

VENTILATION SYSTEMS

Many well constructed homes require some sort of device to provide a source of clean air for the occupants and remove contaminated air (laden with excess moisture, VOC's,) Our preference is the use of a heat recovery ventilation system to provide a balanced air exchange. The balance exchange ensures that we are not creating any positive or negative pressures that may draw undesirable contaminants from other locations (sub-grade gasses, dirt, pollen...) the ventilation systems can either use their own duct systems or connect to a home's existing heating / cooling duct distribution system. Well-insulated homes in cold climates require ventilation to prevent mold growth, sick building syndrome, and promote a healthy environment.

THE BENEFIT OF CONSTANT FAN OPERATION

Adding any of the above noted systems to your home's comfort system will not reap the maximum benefits UNLESS a variable speed, constant fan approach is used. The key element is the ability to slow the fan down to a nearly indiscernable point when we simply want to circulate air without heating or cooling it. If the air is moving, your nifty indoor air quality gadget is working. When the fan stops, so does the operation of anything attached to the ducting. A constant fan cycle also helps even out the highs and lows in room temperatures, providing a more balanced temperature throughout the home.

PAY ATTENTION TO YOUR HOME'S INSULATION SYSTEM

The success of every element I have just reviewed hinges on the emphasis that you put on the quality of the insulation and *vapor barrier* installation. A mere R-value placed on a wall or ceiling cavity is irrelevant if the structure leaks like a sieve. Insist on "energy Star" efficiency and spend extra to provide the best envelope (sealing of the home's shell) you can afford. The money spent here will come back to you several ways.

1. Reduced energy consumption in heating and cooling

2. Reduced drafts and improved comfort

3. Improved moisture retention in the winter

4. Reduced infiltration of impurities that require filtration

5. Smaller HVAC equipment due to the reduced loads (often helps pay for the added costs)

Address the following issues in a tight home:

* Provide adequate bath and laundry ventilation

* Install cold air returns throughout the home to promote good air circulation

* a fresh air exchange (HRV or ERV) system is needed to mechanically ventilate the home

* Carbon monoxide detectors are a must.

Notes:

Construction details

"The sweetness of low price is soon overshadowed by the bitterness of poor quality".

Many homeowners (even architects and builders) never think about conditioning a home when the design is conceived (although I have enjoyed the privilege of working with the rare few who value home HVAC as much as I do!). A budget number for heating and cooling is usually predetermined based on experience, rumors, throwing a dart, a friend's house, or a guess. None of these scientific methods work and the result is usually an under budgeted design and poor performance of the final product. Do your homework BEFORE you build. The design of the home you build or remodel dictates many features of the system you install. Here is a list of points to consider.

1. Sunrooms and massive glass exposures: this wonderful addition to your home adds significant heating and cooling demands and makes temperature zoning more important. These rooms often require heating or cooling when rooms that are more conventional are satisfied and zoning these by themselves is necessary.

2. Tile and concrete floors: a popular and durable choice.

    Have you ever stepped on a 60-degree tile floor in December? Consider the comfort of radiant floor heating if you plan on tiling or using slab on grade construction

3. Finished basement: great use of extra space. Basements need extra care to keep dry and dehumidified. Stand-alone dehumidifiers are ducted into the rooms to provide constant moisture removal and eliminate odors. Ceiling heights are often issues as we try to extend ducting to the floor above while retaining an open ceiling below. I suggest higher basement walls and flush beams to reduce duct twists and turns and eliminate soffits in the basement finished rooms.

4. Post and beam / log homes: sone of the most beautiful homes I have worked in. The high vaulted ceilings, expansive glass (see item 1), and exposed beams/floors leave little room for heating and cooling ducts and piping. Plan on compromises to allow the ducting to fit. Radiant floors work great with these high ceilings

5. Multiple levels: multi floor homes require zoning between each floor to keep the temperatures even and in check

6. Bonus rooms over garages are always candidates for having their own

    thermostat. Rarely do you find a room like this that is comfortable

THE LITTLE DETAILS COUNT

Do your homework and research the finer details that many builders and HVAC contractors usually choose for you. Their motivation is generally to keep the price low and get the job as the low bidder. Presenting prospective sub contractors with a list of questions about their design ideas, equipment, and expertise will send many contractors running for the hills. The last thing they need is a homeowner with more knowledge than they have. If your contractor does not present ideas and options to improve your home's comfort, safety, and efficiency: find a more qualified company to work with. The "recipe" chapter will provide you with an armful of ammunition and test the ability of your HVAC contractor. Successful heating and cooling is not just a factor of installing a furnace or boiler, but the assembly of a multitude of components that should work in unison to provide you with a comfort level that will make you wish you had a home office!

GET THE FACTS

Insist on details from your contractor that specifically outlines the design plan and equipment specifications. I see many homeowners jump at systems because the number met the budget and they never really understood how the system worked or what they were getting. The FIRST step is to review the heating and cooling loads of the house. Without this calculation, how do you know how large (or small) a system is needed to properly condition the space. There is NO level of experience that takes the place of this step. I recently reviewed a geothermal installation that never had this performed before the homeowner dumped close to $50,000 into a new hvac system. Immediately he realized an issue as the home's temperatures soared during the cooling months and ran on electric backup (trust me, you wouldn't want his utility bill) for a large part of the winter. My analysis confirmed that one part of the system was close to 3 tons (36000 btu) short- strange math as the contractor only provided 3 tons in the initial installation. Now the battle begins over who pays for the next $20000 in work to provide a correctly sized design. An important part of any quote is an equipment list with PRODUCT OR MODEL NUMBERS. Many companies make many models of furnaces, boilers, and air conditioners. To say you are getting a brand X air conditioner means nothing, even if it is a major brand name. A name alone does not assure quality or efficiency. Insist on model information and brochures to allow you to research the offering. Manufacturer's websites are also valuable resources to check out equipment. The sites often feature dealer locator sections to see if your contractor is a registered installer for the product being offered. (dealer locators DO NOT assure competence. The contractor noted in the geothermal debacle above was found via dealer locator. Just make sure all of the proper steps are taken BEFORE taking the plunge).

Request an equipment specification page with all equipment listed with model

numbers. Manufacturer's brochures should also be provided for your review

Use the Internet to your advantage

Many contractors are not up to date
on the current state of the HVAC
industry's advances

Important questions and facts that need answers before a contract is signed

1. Insist on seeing a copy of your HVAC contractor's heating and cooling calculations. This will assure you that they have gone beyond the standard "let's put in the biggest system that we can fit through the mechanical room" sizing methodology many inexperienced contractors use. There is a science and engineering base behind the capacity of the system. **This is necessary on every project!**

2. Get an equipment specification list with model numbers

3. Get a signed contract with as much data as you can get. Make sure specific details are covered—don't accept verbal commitments

4. Ask all equipment guides to be collected and presented to you as part of the project wrap up

5. What about maintaining my new system? Ask your contractor to recommend frequency of service information and costs.

6. Are extended warranties available on equipment? This is money in the bank on furnaces, boilers, and air conditioners.

7. Liability and worker's comp insurance on all employees on site. Ask for proof of coverage to protect yourself in the event of an accident. Many contractors beat the competition by hiring "under the table" helpers. This allows their overhead to

be lower as their costs are reduced. This leaves you open to lawsuit if injuries occur.

8. Ask for referrals to check up on the quality of the perspective contractor's work and reputation

9. Ask for sealed combustion boilers and furnaces whenever possible. Most oil burning appliances do not offer this option. The sealed combustion feature reduces your home's heat loss in the winter by eliminating much of the infiltration (air leaking in).

10. Get a time table of the projected work duration

11. Point out areas of the home that present comfort concerns (sunrooms, conservatories, bonus rooms over garages, additions that have many exterior sides) and see how the contractor plans on designing around these issues. The stuttering and stammering is a good indication that you need to look elsewhere.

> Demand detailed reviews of all phases of the project to assure all parties are on the same page and all bases are covered

Notes:

Duct design details

A duct system is necessary to completely address the "four factors of comfort" in a home. I have been in many homes that are one hundred percent radiant and, aside from the floors being super comfortable, several things are missing. The home tends to be dry in the winter. And, watch out, when the summer heat kicks in. Do not get me wrong, I live in a house that has many heated floors and would not EVER install tile without it. I even love my 80-degree hardwood floors; it just does not get any better than that! The completeness of your home conditioning system relies on an assembly of components that act as one to deliver seamless comfort.

Many individuals do not want to consider heating their homes with a duct system. I have been in many homes where I could not agree more. Oversized furnaces blast heat into the ducts and the noise makes your hair stand on end. Those systems need an exorcism to remove the demons that drive one crazy every time the thermostat brings them to life. Many of the non-believers have lived in a home with possessed heating systems and cannot imagine anything else attached to their ducting. But there are numerous ways to add ducted comfort to a home. Modulating gas furnaces and variable speed air handlers add comfort that set a new level of perfection to a ducted home conditioning system. Of course, not every system is a vision of perfection. Even great equipment attached to improper ducting will evoke the demons once again. Duct design must start with the basics, a thorough heat gain and loss calculation followed by a well thought out design. The airflow calculations can be complex, especially in homes requiring zoning. Many installers skip the complex duct assembly and just add extra equipment. This may do the job, but at the sacrifice of efficiency, longevity and value. Maintenance costs rise as equipment ages and enters the golden zone (this is the point where it breaks often and takes all of your gold). Another downfall of zoning with equipment is the cost of adding the "gadgets" used to improve air quality. I would much rather pay for one humidifier, ventilator, and hepa filer than pay for two sets. If the cooling load is less than 5 tons, the home can be safely conditioned with a single system. Anything more is design excess.

The overview below is a breakdown of the necessary steps needed for a complete and effective duct installation. Just like the "four factors of comfort", skipping one can lead to disaster.

Duct installation steps and procedures

1.   Seal the duct system: It is a rarity to see a properly sealed duct system in todays "rush rush / cheap cheap" market. Let us look at the scenario a little closer.   Air is heated or cooled in your air handler and then blown into your home via a duct system. This side of the duct system is under positive pressure due to the fan's power. Any crack or separation in the duct will allow the air that you just paid to condition, to leak to the outside of the ducting where it is wasted. The opposite is true on the cold air return. This side of the ducting is under negative pressure from the fan sucking the air from the home. The cracks on this side of the duct system will allow the fan to draw air in from the space around the duct and allow it to enter the home. This can be from a basement, attic, crawlspace, wall... definitely not advised when we are trying to improve the air quality and reduce the heating and cooling costs . The air leaving the ducts can also cause pressure imbalances in the home, as the balanced air exchange is lost.

2.   Insulate the duct system in all unconditioned: use a good quality insulation to add as much R-value to the duct as possible. This also should provide a vapor barrier to keep moisture away from the duct and help prevent condensation. Many new duct wraps are available to replace the traditional fiberglass/vinyl covering. Foam sheet wrap adds durability, neatness, and added R-value to improve the system's effectiveness. The duct system should be kept low in

attics to allow the insulators to add extra insulation over the entire trunk and branch system. This extra r-value will keep the ducts from absorbing the heat or cold from the attic space, saving you money every day of the year.

3. Design the duct system properly: Make sure a professional does this. An improperly sized duct system will be noisy and provide uneven airflows to the residence resulting in hot and cold spots. Once again, insist on a heat loss calculation that shows room-by-room values and subsequent duct sizing. The installation of the ducting is not something that can be left to rule of thumb generic formulas and guesswork.

4. Watch for excessive use of flexible duct: If it is sized and installed properly, flexible duct branches are fine. This is one of the most abused products in the duct trade. Excessive lengths and turns will reduce the air volumes from the grills to a mere trickle and possibly create even larger air flow issues withing the furnace of air handler (damaged heat exchangers, frozen coils in the AC mode)

5. Zoning a residence: if your home requires less than 5 tons (60000) btu) of cooling, there is no reason that a single air handler cannot condition the house. Two story homes must have two thermostats to provide individual control of the levels. Many contractors will immediately jump to a 2-furnace/2-condenser system when multi level installations are required. Installing two air systems doubles your annual maintenance and long term repair bills and puts an air system in the attic, an unconditioned space and the worst location for an expensive piece of home conditioning equipment. This also doubles the wiring requirements, gas piping... adding even more cost to the system. Installing one system in the basement and controlling its operation with a mechanical damper

/ thermostat control and use a single furnace / condenser to heat and cool the home. This requires a substantial duct chase to the attic, but the lost space is well worth the sacrifice.

Choosing a furnace or air handler

1. *Oil fired furnaces:* this is an option I try to lead EVERY homeowner away from. When a client talks oil and ducting, I steer them towards the hydro-air option (I steer everyone in this direction whenever I can) using an oil fired boiler and hydronic air handler. Please review the hydro-air review on page 17 and 18 for a refresher.

2. *Gas furnaces:* if the added cost of hydro-air is not in the budget and gas is available, there are some terrific gas furnaces available today that provide impressive results. Many of today's furnaces (LP and natural gas) are equipped with two stage or modulating gas valves that actually make one furnace two (or more). When we size your home's furnace, we base the equipment on what we call "design temperature". This is the worst-case scenario for your climate (example: -10 outside and maintain 70F inside). Whenever the temperature creeps above −10F, I have sold you a furnace that is too large for your home and as the outside temperature rises. Wouldn't it be great if the furnace could resize itself and adjust to the reduction in your home's heat loss? Two stage and modulation furnaces do just that. They start on a reduced setting (and use a

smaller gas flame) and run on the low output for as long as possible, increasing the flame size only when the thermostat realizes it is not getting warm fast enough. This multi stage operation will lengthen the fan cycles and create a more even temperature throughout the home. The increased fan operation also allows the humidifier, filters... to add their benefits to the home (remember when the fan is off– those air quality components sit dormant). The next step up in furnace design marries the multi stage gas valve operation with a variable speed blower (ECM blower). This is the ultimate in gas furnace technology, providing quiet operation, enhanced comfort, and maximum dehumidification during the summer cooling season

3. *Air handlers:* this component is the staple of the hydro-air system or when a cooling only system is desired. It is simply a cabinet with a blower and heating and/or cooling coil inside. It is the means of moving the air through the ducts and adding the heating or cooling capacity to the system. These can be simple single speed boxes that come on and off with a call from the thermostat, or more advanced systems with ECM motors that vary the fan speed to enhance the home's comfort. The success of our systems and the comfort our clients enjoy is our extensive use of variable speed air handlers coupled to a hot water boiler for heating and high efficiency air conditioning condensing unit for cooling.

4. *Manufacturers and warranties:* insist on quality equipment with extended warranties. The investment in an extended parts and labor warranty is money in the bank.

The hydro air advantage: the best of both worlds

If you are looking for a design that addresses every aspect of home comfort, this is

your only choice. Many of the "four factors of comfort" require a duct system to

properly condition the home. Direct fired duct systems (using a gas or oil warm air

furnace) work great in many applications but can fall short when the entire HVAC

package is reviewed. I prefer this approach on almost every project.

The core of the system:

The heart of the hydro-air system is the boiler that provides the heating capacity.

The goal is to take a hot water boiler and use this appliance to provide the home's

heat and add the benefit of domestic hot water. This is accomplished by using an

---

indirect water heater (a large thermos that sits next to the boiler and stores water that is heated for bathing...). These tanks have coils inside of them that allow the boiler to circulate hot water to indirectly heat the stored water. This "heat exchanger" allows the use one fire to satisfy the home's heating and hot water demands. This reduces maintenance and increases efficiency. The boiler allows us to add additional hydronic (hot water) heating sources to add comfort to the home.

The duct system:

Now that we have a boiler to provide the raw energy for heating, we need a duct system to carry the heated air throughout the home. The key feature of this system is the air handler. These appliances have heating and cooling coils inside that allow the blower cabinets to do double duty. The boiler circulates hot water through the coil in the winter for heating and the exterior air conditioner sends refrigerant through another coil in the summer for cooling. The best models feature variable speed fans that change speed based on demand. When connected to a properly sized and installed duct system, these are super quiet and provide even temperatures. The variable speed fan allows constant fan use to allow the air filters and humidifiers to do their jobs (when the air is not moving, your investment in air quality is not helping!)

Here is where it gets interesting. You have a duct system for general heating and cooling. (If you have a standard furnace, the excitement stops here) The boiler allows us to add ANY feature that requires heated water for operation. The list below provides an impressive review of options that make a house a home.

1. Radiant floor heating (very cool!)

2. Pool and spa heating

3. Snow melted sidewalks and driveways

4. Heated towel warmers for added comfort

    in the bathrooms

5. Radiant wall panels for added comfort in

    problem rooms

6. Standard baseboard hot water zones

7. Radiant heated garage or basement floors

8. Weather responsive control over the boiler temperature to add extra comfort and efficiency (see the "circulation is the key to comfort" essay)

9. Reduced maintenance due to the use of a single boiler in place of multiple furnaces

10. Ducted central air conditioning

11. Indoor air quality: the use of a duct system provides an avenue to add a multitude of indoor air quality improving devices.

Extending the comfort of circulating air

I hope that you can see the benefits of combining a boiler with an air handler for air movement. It allows us to address all of the factors of comfort to create the ideal indoor environment. Constant air movement will keep the air quality fresh by allowing the individual filters, humidifiers, and ventilators to do their tasks. The constant fan will also allow the room temperatures to stay fairly even. This strategy works well with rooms with lots of exposure. Central thermostat locations tend to make the rooms around the perimeter cool down too much before the thermostat turns the heat back on. One drawback of constant fan operation is the possibility of a cold draft caused by the air movement. Poorly designed duct systems also tend to be noisy and the thought of that nuisance continuously is less than desirable. If you are installing a new system, consider the use of an air handler with a variable speed fan. The variable speed (ECM) motor will reduce the noise and drafts as the

fan speed is reduced during fan only operation. Many systems also allow the user to

program the air handler or thermostat to adjust the heating fan lower than the

cooling speed. The variable speed capability also has a dramatic effect on the

system's dehumidification ability during cooling season. The fan speed is reduced

during the first several minutes of operation, wringing the water from the air with a

cold air conditioning coil. As the speed ramps up, the coil continues to dehumidify

as it cools the air. The ramping up and down also quiets the duct system and

eliminates the annoying noises that the air movement can create.

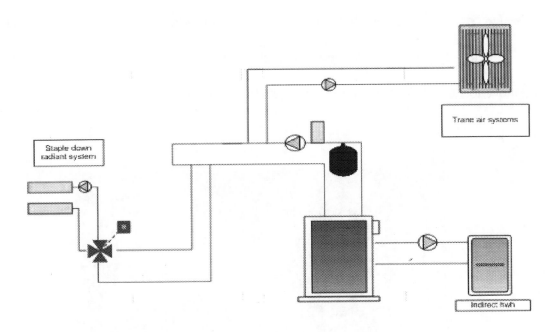

Simple concept sketch showing a boiler's job in a hydroair design. As you can see, one heating unit can add a lot of value to a system.

The drawing above is an example of a simple hydro air design, one we have used countless times in one version or another. The sketch shows the boiler with its satellite heat demands and associated piping. The versatility of this design often makes me wonder why this approach is not used on every job. Initial cost seems to be the number one drawback to these installations. There never seems to be enough money allotted for even basic systems, never mind something more advanced. With this overwhelming obstacle before us, education and an understanding of the exact short and long term costs are essential.

Don't make this mistake!

The key to every design lies in the duct design and installation. No matter how good the equipment is, the comfort it delivers is linked to the duct system. Improperly sized ducts cannot allow the HVAC appliances to work to their true potential.

An example of a low budget ducted air design

The duct system above is a great example of a collection of don'ts. The duct system has been modified for years and new rooms lead to new extension ducts. The new furnace was attached to the duct system without regard for proper distribution. Ducts were removed and never capped; gaping holes where the seams have split... This system required total removal before the home's comfort could be resurrected. We have seen new homes with duct systems in a similar condition. Ducts left open above insulated ceilings, massive flex duct runs (with little or no air coming from the ducts), unsealed distribution systems leaking conditioned air into attics and crawl spaces.

Everyone associates comfort with the equipment that is installed, assuming any problems are rooted there. Buying world-class equipment and connecting it to a subpar duct system will not deliver the savings or comfort you had hoped for and potentially damage the equipment. Every spring I get calls from homeowners interested in adding air conditioning to an "ac ready" system. When I get there and look at the ducts, my first response is that the duct system is insufficient for the application and a major duct renovation would need to take place at the same time. Several of these homes were only a few years old and you can only imagine the look I get when I condemn a new HVAC system's ductwork.

These two high efficiency gas boilers provide the heating capacity for the home's domestic hot water demands and all space heating. Four remote air handlers serve the duct system for heating and cooling. Each air handler is fitted with a hot water heating coil that receives heated water from these boilers. This system replaced a 20plus year old geothermal system that was costing over $20,000 per year in electrical use to condition the home!

NOTES

Hydronic system design

What is a "hydronic system"?

This is simply an industry term for a system with a hot water boiler and some sort of heat emitters that use circulated hot water from the boiler. This is the foundation of the hydro-air system and strict hydronic systems using only radiators or radiant heat. Hydronic heat actually is a minority in the US heating market, seeing the most use in the cold Northern climates where cooling is not a demand factor. In these areas, radiant floors, baseboard, hydro-air, and panel radiators add unsurpassed comfort to those lucky enough to own a system.

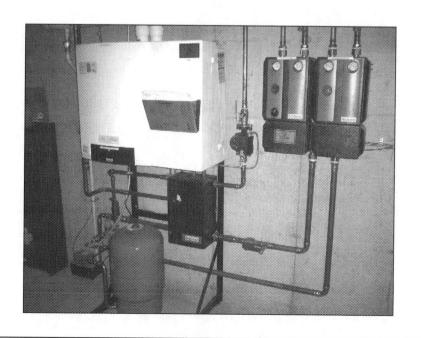

The drawing below depicts the versatility of a boiler-based system. We have shown a single boiler providing five services for the home. This enables the full, year round use of the boiler and reduced annual maintenance of direct-fired appliances.

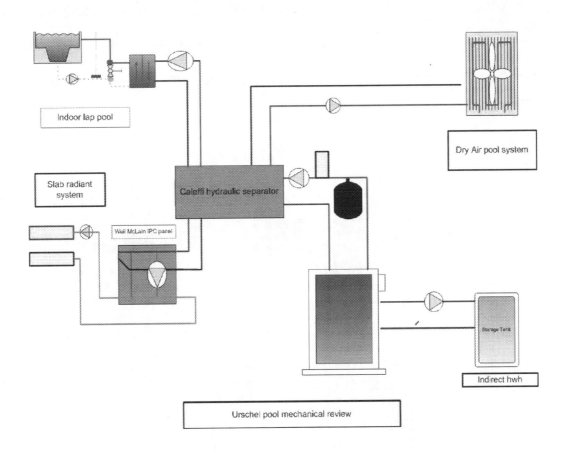

Indoor lap pool

Dry Air pool system

Slab radiant system

Caleffi hydraulic separator

Weil McLain IPC panel

Storage Tank

Indirect hwh

Urschel pool mechanical review

The versatility of hot water heating is attractive to those individuals living in cold climates. The comfort delivered by radiant floor and wall panels is unattainable with any other kind of heating. The ability to provide multiple zones of temperature control with hydronic heat is another fantastic benefit. The European radiator influx to the United States has provided consumers with a heat emitter superior to our standard hot water baseboard. The panels are decorative, durable, and provide a warm blanket of comfort in the rooms they serve. The radiator pictured in this photo below features a non-electric or self-contained thermostatic actuator. This device is part of the piping system, varying the water flow through the radiator as the room heats up and cools down. This device allows multiple zones of control without the need for wiring and conventional thermostats. These radiator-mounted zone control heads have numbered scales that allow the room occupants to increase or decrease the radiator output.

Radiant floor heating is another magical addition to hot water heating. This concept has been popular in Europe for decades and has seen a rise in popularity in the States over the past 15 years. The principal behind radiant heating is simple; water is circulated through the floor, creating a huge radiator. To safely provide this comfort, the fluid temperature within the radiant floor pipes must be tempered through a mixing device. Water that is too hot can damage floor surfaces and careful attention to this crucial design detail is imperative. Several options exist for this purpose, ranging from basic fixed tempetaure devices to three and four way mixing valves with motors and variable speed injection mixing pumps. The advantage of more advanced stations is their ability to adjust the water temperature in the floor to the room's demand. Most of these advanced micro processor based controls use the outdoor temperature as a reference point to determine how hot the water in the floor system needs to be. The colder the weather, the hotter the water. Many use the indoor temperature as another reference point to eliminate over-conditioning and improve efficiency and comfort. The principal is to maintain the water JUST warm enough to keep the room comfortable. The circulator moves water through the pipes as long as the outdoor temperature is low enough and produces the wonder of a weather-responsive control system. I use the essay below to paint a picture in our clients' minds of what the addition of this kind of control system will do for their homes. It converts a complex assembly of electrical controls into a concept that every layman can comprehend.

THE BENEFITS OF WEATHER RESPONSIVE CONTROL

Allow your heating system to keep up with your home's heat loss with weather responsive controls

The heating system is the heart of our homes; and like our hearts, its operation determines how we exist. Can you imagine the repercussions if our heart operated like most heating systems? What do you suppose would happen if blood flow was established as each section of our bodies needed nourishment instead of providing a constant flow of essentials throughout our entire body as it does now? We would feel hot and cold because the temperature in our extremities would never stay constant. Maintaining a constant balance within our bodies would be difficult because of the varying flows, and the healthy tissue would suffer from the inconsistent pattern. Do these symptoms sound familiar? I diagnose more unhealthy heating systems in a year than I thought possible. Many of the problems can be resolved by following the role model on which we all rely—our hearts.

It is the constant flow of blood that maintains a stable body temperature, a feat that would be impossible without the blood's constant circulation through our veins and arteries. Now picture your heating system in place of your body. The vessels are replaced by pipes, and the picture starts to evolve. Can you imagine the comfort we would enjoy if our homes maintained a stable temperature plateau like our bodies?

Well—thank your lucky stars—we have the prescription for the illness that makes many otherwise happy homes about as comfortable as a locker in the morgue. By allowing the boiler to circulate water through the radiators at a constant rate, we can eliminate the ups and downs we feel as our heating system cycles on and off trying to *catch up* to the heat loss. If we allow the system to *keep up* with the heat loss, the peaks and valleys will be erased; and your home's *cold* will be cured.

The function that allows us to provide your system with the ability to *keep up* is referred to as *weather responsive control.* By using an outdoor temperature sensor to provide feedback to the boiler, we can change the delivery temperature of the boiler to match the heat loss of the home. As the temperature outside changes, so does the temperature of the water within the hydronic system. This constant monitoring of the environment allows us to keep up with the heat loss of your home. As the temperature gets colder outside, the heat loss of your home increases. The sensor tells the boiler to fire a little longer and raise the temperature a little. *Keeping up* provides the comfort and fuel efficiency that makes this kind of of control system unique.

In addition, by providing a constant flow, this system produces other design possibilities. Non-electric valves can be used on individual radiators to provide zone control as needed. In this way, each radiator can become its own zone to provide even further comfort and economy of operation. We have designed many systems in which the owner has over twenty zones of temperature control. The cost of using these valves is minuscule when compared with conventional installation and maintenance costs.

Radiant floor heating is another fantastic function of constant circulation. Can you imagine your cold tile or hardwood floors at a constant 80°? Cozy? You bet! Nothing beats the comfort we can provide with a little tubing under the floor.

We can also extend the lifespan of the system's components by changing the water temperature in which they work. Most applications can use 150° water or less to comfortably heat your home during 75 percent of the year . The reduction in thermal stress will also save you big bucks in repairs.

Comfort, economy, and reduced maintenance—the benefits of adding weather responsive control to your home's hot water boiler

The basics of radiant heating installations

1. The MOST important design feature: use a smart control system that will provide weather responsive control over the radiant flooring. This will allow the room temperatures to remain even, without the typical ups and downs seen with most stop/start systems. The best option for residences that feature large radiant expanses is a system that manages the radiant temperature two ways, using the outdoor temperature first and the room temperature as feedback. If the room overheats, the radiant system backs down and circulates cooler water. When the outdoor temperature dips and the room cools, the mixing device will allow hotter water to enter the system and sustain the room temperature.

2. Do a heat loss evaluation of the space: radiant floor heating is a subtle source of comfort and requires careful attention when designing a system. Rooms with large glass expanses and multiple exposures can push the radiant heating to the limit and fall short during periods of extreme cold.

3. Installation methods have varying heat outputs: depending on what kind of radiant installation your project calls for, the output of the floor will vary. An installation featuring tubing embedded in concrete with a water temperature of 100 F will provide 28 BTU / sq ft. A staple up system (tubing attached to the bottom of the floor surface) under two layers of wood (plywood and hardwood floor) with the same 100F delivery water temperatures will provide 12 BTU /sq. ft. Knowing the composition of your floors is vital to a successful design and

matching these outputs to your heat loss calculations will assure satisfactory results.

4. The radiant system MUST have insulation beneath the tubing whether the tubing is installed in concrete or under/over a traditionally framed sub floor. Insulation is added to drive the energy in the desired direction. When radiant is installed in concrete slabs, make sure that the slab is isolated from the foundation by a layer of insulating foam. This will keep the cold foundation from siphoning the heat from the slab

5. Seek professional assistance: The do-it-yourself shows and websites try to make radiant systems simpler than they are. Proper design and installation takes years of training and education and Internet "home radiant installation kits" will cost you more in the long run. If you are determined to do some of the work yourself, find a consultant to help with the design and planning. This expense will pay off in the end.

The keys to succesful radiant installations under hardwood floors

Basic Design Principles

- Heat transfer plates are needed in every staple up or staple down application. Without cement as a heat transfer medium, something is needed to help extract the energy from the tubing and spread it into the floor. Many installers simply nail the tubing up in the floor bays and insulate below. Hotter water temperatures are required and the efficiency factor is lost. These plates can be pressed aluminum flashing, extruded aluminum rails, or hybrid wood/aluminum systems.

- Weather responsive control of the floor heating system is a important design element that ensures that the floor will remain at the lowest possible temperature for as much of the heating season as possible.

- Use the maximum insulation below the tubing to ensure that the maximum Btu output will be directed at the living space.

- Balanced flows within the radiant system can be established either through the use of similar circuit lengths or, preferably, through the use of flow control devices on the header.

- Use of an accurate mixing/tempering device for the floor heating circuit with high limit safety.

- Maximum zoning enables rooms or groups of rooms with similar heat loss characteristics to be isolated.

- Separation of floor heating circuits that have different heat transfer capacities (i.e. slab pours, carpet, hardwood).

- Staple down applications are best in dry (non concrete) applications as the greatist heat extraction is realized due to the reduced mass (layers of wood floor) the radiant energy must pass through.

- Be VERY careful installing wood over concrete floors. Concrete can take months to cure (and dry) properly and the water will still exit the mass for quite some time. Immediately installing wood over this surface will allow the fibers to act as a sponge and absorb the vapor leaving the cement. This can cause swelling, heaving, and other unsightly and damaging abnormalities in the floor. Proper vapor barriers may be needed but be sure to give the cement a chance to properly condition. Testing tools are available to check the moisture levels of the cement before flooring is added.

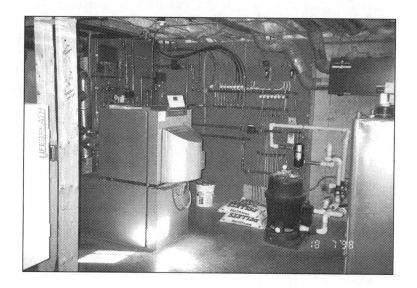

Jobsite Requirements

- Stable relative humidity of the structure should be ensured

- Drywall should be completed before hardwood is delivered.

- If gypsum is installed (this mixture is typically poured over the wood subfloor), the moisture content of the slab must be established before the floor is laid or delivered

- Use of proper vapor barriers as required.

- Ensure that all of the moisture emitting trade tasks are completed before hardwood is allowed on site.

- Verify that the moisture content of the flooring is within the manufacturer's recommendations before installation.

Selection of Hardwood

- Narrow-width flooring is preferable. Ideally 3 inch maximum.

- Wood species should be investigated due to varying properties for each kind of.

- Attention should be given to the method of cutting the flooring. Quartersawn, flatsawn, riftsawn cuts will all lend different expansion/contraction properties to each wood species.

- If you are looking to maintain a tight floor with natural wood, stick to narrow strip flooring and use quarter sawn material.

Overview

Hardwood flooring over radiant heat is not a problematic issue when the factors listed above are taken into consideration. The critical hydronic issue is low temperature distribution and weather-responsive controls.

The largest overall issue is control of the climate in which the hardwood flooring is installed. Humidity is the largest single issue for preserving the integrity of a hardwood floor. Assuring that the flooring is installed in a safe environment in relation to the moisture content of the subfloors and building interior is necessary to ensure long-term success of any hardwood installation.

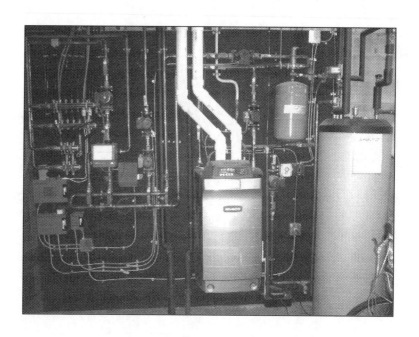

Basic boiler installation details

1. Use an advanced air removal system to keep the air out of the system. Many manufacturers offer products to improve this vital facet of the design

2. The engineering of the boiler's piping system is crucial to longevity and trouble free operation. Put the pumps on the supply and pump away from the expansion tank.

3. Avoid power venting / direct venting oil whenever possible. Oil is not a clean fuel source and making the mistake of sidewall venting an oil boiler will be one of your first regrets. Manufacturers do make attachments to make this feat possible but most are short lived and costly to replace. The sidewall venting can also lead to the staining of your home's siding (from soot blown back on the house by prevailing winds)

4. ALWAYS buy the most efficient boiler that your budget allows. You are married to your fuel bill as long as you own the home and the investment in superior equipment will pay for itself over time.

5. Add extended parts and labor warranties to your new boiler if available. The best value is found in a 10 year parts and labor coverage

6. Make sure the installer uses ball valve isolation valves on all supply and return lines to facilitate service and purging

7. Bring combustion air in to the appliance. A boiler can use thousands of cubic feet of air a day to sustain combustion during the winter months. This air will be

sucked in through every crack in your home and increase the infiltration losses of your residence (and increase your fuel consumption).

NOTES

Alternative energy solutions

Solar thermal alternative energy concepts

Allowing the sun's energy The rising cost of fuel and the long term ramifications **to remain untapped is like** of pollution makes the concept behind solar energy a **watching dollar bills fall to** fascinating avenue to explore. If your home features a **the earth and disappear** strong southern exposure, adding the benefit of radiant heating may be a viable alternative for boosting your domestic hot water production or even acting as a space-heating source for your home. As long as the sun is shining, there is energy waiting for harvesting. If I look closely enough on a bright, sunny day, I can almost see dollars floating from the sky (don't we all wish) and disappearing into the earth below my feet. If you think about it, that is what is actually happening if we are not extracting this energy that is so abundant. Solar can successfully provide domestic hot water for our homes throughout most of the year (as long as the sun is out). The advances in solar collection panels make the task even more efficient as vacuum tube designs are fast replacing the flat panels of the 70's . The vacuum tube technology has many advantages, the best being the increased production of the thermal harvesting (energy absorbed by the panels). I was at the home of the panel system on a brisk (25 F.), sunny day in January. The air was cold, but the panel sensor was showing 145 degrees Fahrenheit! With temperatures available of this magnitude, low temperature space heating could also be possible. Radiant floor heating, especially in concrete, is a great match for solar boosting. As the design expectations increase (adding space heating...), the engineering principles follow suit. Adding panels for winter service now leaves us with a massively oversized summer system when only a small domestic storage tank is used. Dealing with this excess is vital for proper performance and trouble free operation. A dump zone (somewhere to get rid of the excess energy) is often used or, a drain back style flat panel system is installed. The drain back is ideal in northern climates (where freezing is an issue) and in oversized designs as it allows the panel fluid to escape to a storage tank when the control ceases solar harvesting. This can be very complex so be sure to bring an installer on board with experience in varying design applications.

Solar domestic hot water heating

This use of solar energy is the most basic and works in every system with proper panel orientation. One simply collects the energy on the roof panel system and pumps the heated water through a heat exchanger / storage tank where it will be stored. The solar control panel looks at the roof panel temperature in relation to the storage tank temperature. When the roof is hotter than the tank core, the circulating pump turns on and the sun's bounty is harvested. We recommend large storage vessels as the capacity of smaller tanks would quickly satisfy the sensor system and the energy collection would stop, leaving those dollar bills falling to the earth. ALWAYS install a good quality mixing device on the hot water outlet from the water tank to your faucets. This will prevent scalding and burn hazards presented by the superheated tank water. The water going to your faucets should never be hotter than 130 F. The concept drawing below provides the basic design principles for a closed loop system requiring an antifreeze charge (to prevent the lines from freezing during the winter). Please refer to the manufacturer's design details as I have omitted many components for simplicity sake.

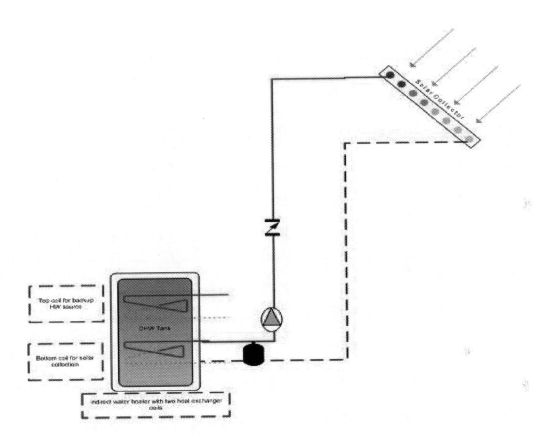

basic overview of a solar system heating a home's domestic hot water

Advanced solar harvesting: adding space heating

If your home's comfort system features radiant floor heating, using the sun's energy to augment the heating demands may be an alternative to 100% fossil fuel use. The panel array size on the roof will need to be increased, as more collectors are required for the new demand. The best way to maximize the sun's potential is to, once again, harvest and store the energy. We would use a buffer tank(s) for storage and pump the water from the tank to our home's heating system upon demand. Slab on grade radiant is the best use of the energy as we can use low water temperatures to condition the home. Slab on grade systems can operate comfortable with water less than 100 F, well within the reach of a solar collection system. The key is storage, grab the energy while it is available and put it away for use when the home cools down. At some point, the cost of storage and additional panels outweighs the benefits and a happy medium must be reached (these components can be expensive and there is a point where the investment sees no return).

Geothermal heating and cooling:

Geothermal systems have enjoyed a resurgence of interest as fuel prices creep upwards and the "green" energy revolution gains momentum. New tax incentives and rebates have added interest to an avenue often cited as too expensive. Expensive? Absolutely. Worth the investment? It depends on the circumstance. The beauty of having the ability to choose from all of the items in this book to condition a client's home is the freedom to choose the right system for the application. Many contractors force a system design because that is all they have, right or wrong, they need to sell something to survive.

As a geothermal contractor, I solely focus on closed loop systems. There may be viable open loop applications, but they are rare in my eyes. I see most open loop applications gaining popularity for budget pleasing reasons only. Closed loop can be costly and is viewed as more of an investment in long-term savings and "green" prosperity. My experience with open loop has been very unsettling (as a consultant only- reviewing the systems after they have fallen to pieces) and hence my desire to avoid the design altogether. There are ways open loop designs work well, just avoid that option because of cost and always work with an individual with open loop experience and proven success.

What is looping?

Looping is the portion of the design that is exposed to the soil, providing the energy transfer. Using vertical wells or horizontal trenches, an extensive plastic piping system is used as a means of circulating a heat transfer fluid. This medium (usually water and antifreeze) is pumped through these pipes and back to a machine that does the geothermal "magic". In the summer, the system extracts heat from the home and deposits it into the ground, the reverse in the winter. The key is the geothermal heat pump everything is connected to. This electrically operated device uses a compressor to manipulate the temperatures and create heating or cooling. These systems can operate at efficiencies 300 to 400 percent greater than conventional fossil fuel burning designs. The kicker is the upfront investment; you are paying for years of savings in the beginning and enjoying the reduced operating costs and environmental benefits for decades to come.

Is geothermal worth the investment?

In the correct application, without a doubt. Geothermal designs are very flexible, similar to the hydro air concept. We can make warm air, warm floors, heat a pool, produce hot water for bathing, and even melt snow. The designs are a little more restricted, as only low and medium temperature water / air can be achieved. As long as the home's energy retention capacity matches the low-level energy production of the geothermal, this idea is a homerun.

Planning for the initial cost can be problematic as the HVAC is often decided late in the game, when most of the money is already spoken for. Smaller homes or super insulated dwellings are an easier fit for geothermal. The heat loss / gain calculations are small and the looping requirements are much less. Older homes with poor energy retention (poorly insulated with drafty interiors) are poor candidates for geothermal as the amount of exterior looping would put any homeowner into a state of cardiac arrest. In the northern climates I design in, a cost-cutting trend is to design the geothermal system for a percentage of the total heating load, say 75%. I then utilize a fossil fuel backup heat source to carry the balance of the load and even assist with the domestic hot water production (boiler type design). As usual, the options are endless and drawing on an experienced entity for this phase of construction is vital for success (don't be the guinea pig!). Even if the numbers are daunting, geothermal warrants a consideration in most homes.

NOTES

Allowing our homes to breathe

If you follow our advice and insulate your home properly, using a ventilation system is mandatory to provide a balanced air exchange in the home and a source of fresh air.   Many people insist that homes should not be constructed so tight that ventilation must be added– "give the house a chance to breathe!" I agree with the "chance to breathe" portion of that statement, but let's do it mechanically, with a controlled air exchange that brings in fresh air, captures heat from the air being exhausted to the outside, and preheats the cold, incoming air . By ventilating mechanically, we can say how much air and when– allowing maximum control over the indoor environment. This balanced flow of air keeps the house under neutral pressure as we exhaust the same volume of air as we take in. Negative pressure within a structure can allow radon, allergens, fumes and other undesirables into our environment. The illustration below is a basic overview of the requirements for a home ventilation system. The ventilation appliance is installed and two ducts extend to the exterior. One is a fresh air intake and the second is the exhaust air from the home. The two air streams cross, but do not mix, in the ventilator. An internal heat exchanger preheats the incoming air with the warm air discharged to the exterior of the home. From this point, the preheated, clean air is ducted to the living space.

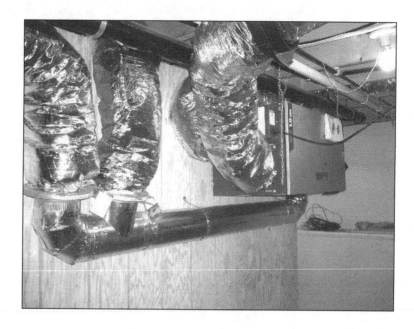

Integrating ventilation into your home's comfort system

If your home's conditioning system features a central duct system, the ventilation system's installation requirements are reduced dramatically. We no longer need to create a distribution system for the home; we can simply use the HVAC systems ductwork and fan power. The ventilation system would tap into the cold air return, 6 to 8 ft between tappings and sample air for the exchange. We would still need the fresh air supply and exhaust to the exterior, but eliminating the extensive interior ducting will save considerable capital. The ventilator can be interfaced with the air system's operation (allowing it to run when the furnace operates) or better yet, a constant fan system would simply allow the ventilation system to operate as its control demands, on a timed call for ventilation or a high humidity signal.

Heat recovery ventilation systems (HRV's) can also serve as bathroom exhaust systems for the home. The bathrooms can be fitted with exhaust ducts and a timer switch that pulls in the ventilator when exhaust is needed. Careful attention to duct design and system capacity is required as most ventilation systems have small fans and would not provide sufficient air movement for multiple baths that are far apart. One note for cold climate use, over ventilation can lead the removal of too much moisture from the home. The better systems have integrated control pads that allow intermittent operation which eliminates over ventilation and gives the operator a choice of settings.

Unique ventilation applications

Occasionally you run across a unique design that requires extra effort to maintain healthy living conditions. Tightly constructed homes are usually the most basic application requiring extra attention to detail. Another function that warrants special care is indoor pool and spas. Once again, each individual application is different and recognizing how and when to introduce extra equipment is the key. Some small pools or swimming trainers can exist with home ventilators alone; some require units that are specialized for pool environment conditioning.

Indoor pool environments are difficult to condition, especially in colder climates. We must be wary of condensation, on visible surfaces and within the structure itself. Maintaining a negative pressure in this case is typically necessary, drawing a slight negative pressure within the room to eliminate moisture and odor (pool chemistry) migration. Positive pressure is the enemy in these environments. Driving moisture outwards can lead to devastating consequences down the road. Maintaining an air stream over the glass surfaces is essential, keeping the glass dry and clear. Dust design is critical here, you must assure the air has the velocity and "throw" to reach all of the glass and maintain quiet operation. Skill and experience at this level is the "make or break" feature of this demanding application.

All of the ducting extends back to the mechanical room, where we house the heart of the system- the pool environment conditioning system. The basic concept behind this massive metal box is a dehumidifier with a fan. The device removes the excess moisture from the air as it passes by, similar to an air conditioner in AC mode. Without a machine with this ability, the relative humidity of the poolroom would rise to dangerous and damaging levels and quickly lead to mold and structural issues. As always, experience is critical with a design as demanding as an indoor pool and success hinges on a well thought out and executed design.

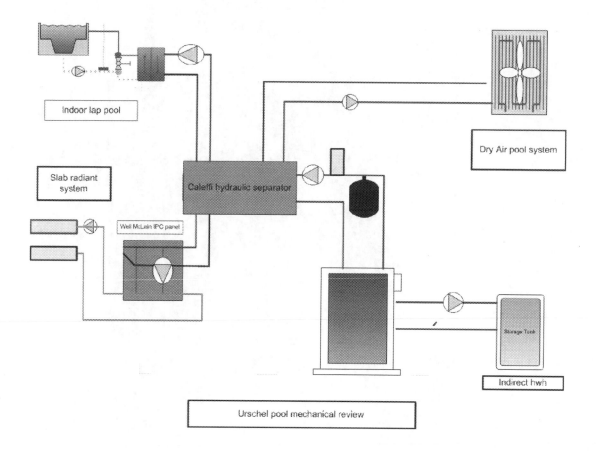

Indoor lap pool

Slab radiant system

Caleffi hydraulic separator

Weil McLain 8PC panel

Dry Air pool system

Storage Tank

Indirect hwh

Urschel pool mechanical review

The sketch above depicts a hydro-air concept of a typical pool conditioning system. I prefer the use of a boiler based system as it allows the use of several hot water based conditioning options. As you can see, we are using a single boiler to heat the pool water, pool room air, radiant floors and hot water for bathing. Clean, simple and easy to maintain- what more could you ask for?

Here is another version of an indoor pool we have worked with. This small bathing pool has a jet that creates a strong current to swim into. These typically need little more than a ventilation system if covered most of the time. We have had great success using a simple heat recovery ventilator and humidistat control. The pool is heated by the home's boiler using a plate-to-plate heat exchanger.

NOTES

Leftovers!

As you can see, cookie cutter designs do not apply to quality home conditioning systems. It takes an artist's refined hand to apply the proper brush strokes to the picture, blending all of the colors into a masterpiece. It is often difficult to resist the magnetic draw of low price, especially after viewing the total cost of a project or renovation. There are few items that can be altered later, heating and cooling is one aspect that must be designed and installed properly from the initial stages. Most of the experience I have gained has been a byproduct of consulting, analyzing installations to see where they went wrong and assessing damage control to set the ship straight again. My first priorities on every project are establishing the heating and cooling system requirements and budgets, addressing any structural issues hindering the installation, and confirming a quality structural envelope (insulation system). The second phase is the education process, making sure everyone understands the demands of the home and how my design fits the application. Many projects are awarded even before the budget is established, solely on an ability to exercise the knowledge that few other HVAC professionals possess. If you are approaching a project as a homeowner, architect or builder, recognizing the demands of your project will help you weed out the "professionals" from the professionals and facilitate choosing your HVAC contractor. One of the key elements to succeeding in this field is the ability to adapt a design and think outside of the box.